trotman

REAL life
GUIDES

MANUFACTURING & PRODUCT DESIGN

ROGER JONES

Real Life Guide to Manufacturing & Product Design

This first edition published in 2009 by Trotman Publishing, a division of Crimson Publishing Ltd, Westminster House, Kew Road, Richmond, Surrey TW9 2ND

© Trotman Publishing 2009

British Library Cataloguing in Publication Data
A catalogue record for this book is available from the British Library

ISBN: 978-1-84455-195-8

Typeset by RefineCatch Ltd, Bungay, Suffolk

Printed and bound in Italy by LEGO SpA

CONTENTS

ABOUT THE AUTHOR

Roger Jones is a freelance writer who has written extensively on careers topics, education and living abroad. His handbooks include *You Want to Work Where?!*, *Charity and Voluntary Work Uncovered*, and *Real Life Guide: Electrician* – all published by Trotman.

FOREWORD

This *Real Life Guide to Manufacturing & Product Design* offers practical information on every aspect of training for and finding a job in the field. Whether you are just starting out or looking for your next career move this book shows the different entry routes into the industry, gives you an outline of the jobs available, and explains the skills and attributes you need to be successful.

City & Guilds vocational qualifications support learners from pre-entry to professional level and we award over a million certificates every year. Our qualifications meet the latest industry requirements and are recognised by employers worldwide as proof that candidates have the knowledge and skills to get the job done.

We are delighted to be a part of the Trotman *Real Life Guides* series to help raise your awareness of these vocational qualifications – we are confident that they can help you to achieve excellence and quality in whichever field you choose. For more information about the courses City & Guilds offer check out www.cityandguilds.com – get yourself qualified and see what you could do.

City & Guilds

INTRODUCTION

'The workshop of the world' – that is how Britain used to describe itself. Its products were exported all over the globe – cloth, chemicals, machines, pottery, ships, steel among them – manufactured in the great industrial centres of the Midlands, the North West, Yorkshire, the North East, Glasgow, Belfast and South Wales.

The situation is different today. Many of the products we buy – cameras, clothes, computers, electrical goods, to name but a few – are imported from China, Korea, Japan, India and other Asian countries. The countries of Eastern Europe are also expanding their manufacturing capacity and exporting to us. It is all part of a process known as globalisation.

In Britain itself the majority of people no longer work in manufacturing. Nowadays it is the service sector which accounts for the majority of the jobs, and this includes finance, retailing, hospitality (hotels and catering), transport, health and the public services.

The figures tell their own story. In 1978 manufacturing employed nearly 7 million people (29% of the workforce); now the figure is down to 3 million (11%). Manufacturing now represents one-tenth of Britain's economy; 30 years ago it accounted for one-third.

Looking at these figures, you might feel that manufacturing in Britain has no future and is best avoided. But don't be put off. One reason why fewer people are employed in manufacturing is because industry is much more efficient these days and is more reliant on machinery to do all the donkey work. As a result most of the jobs on offer are far more interesting and challenging than they used to be.

Manufacturing is definitely not a dead end. If you want proof of this, consider the following examples.

Have you flown recently? If so, did you notice where the plane's landing gear was made ... or its instrumentation, its wings or its engines? Even if the plane was a Boeing or Airbus, there's a strong possibility that some (or all) of these components were made at a factory in Britain by a firm like BAE Systems, GE Aviation, Rolls-Royce or Messier Dowty. The British aerospace industry is very much alive and turning out hi-tech products that are quite outstanding.

What have you eaten today? Did you have a breakfast cereal, for instance? Did you make yourself a cup of coffee using instant coffee? Perhaps you had a biscuit with it ... or a bar of chocolate? When you felt peckish you may have bought yourself a pork pie or a ready-made sandwich and washed it down with a smoothie. Britain's food processing and drinks manufacturing industry is a world leader.

We continue in this vein. Look at the chair you are sitting on or the table you are sitting at. Where were they made? The odds are that they come from a factory somewhere in Britain. The same goes for the mugs and plates and cutlery you use. The bricks and tiles of the house you live in as well as the windows and the doors – all these are likely to bear the imprint 'Made in Britain'.

There are many other success stories like this which disprove any gloomy suggestions that British manufacturing is in freefall. Did you know, for instance, that three-quarters of the cars manufactured here are exported? Or that since 1990 Britain's manufacturing output has grown by a quarter?

Britain continues to be a manufacturing powerhouse. Admittedly some well-known firms have disappeared in recent years, but despite this the country still boasts around 150,000 manufacturers, their annual turnover is £500 billion and they account for 60% of our export earnings.

So, far from being a dead end, many sectors of manufacturing – from pharmaceuticals to the textile industry – offer a bright future for people with commitment and initiative. If you like practical work, are keen to accept a challenge, are methodical and resourceful and want to have something tangible to show for your efforts, there could be an exciting future for you in manufacturing.

Finally, if you feel manufacturing is for people with little imagination, can you say what Terence Conran, George Stephenson, Mary Quant, James Watt, Alec Issigonis and James Dyson have in common? Answer: They are all designers whose ideas have all resulted in manufactured products.

Design is an essential part of manufacturing. Before a product can be made it has to be designed – whether it is an item of clothing, a vacuum cleaner, a car, a piece of furniture or a battleship. In order to remain competitive, manufacturers need to come up regularly with new designs or improve existing ones.

CHAPTER 1
SUCCESS STORY

DAVID HAYWARD

Designer

David Hayward is a designer/manufacturer who has his own design studio and specialises in high value gift items including pens and jewellery.

'I was educated in a technical school where the curriculum was biased towards art, woodwork and metalwork. I then went on to do a one-year foundation course in art and design, which offered an introduction to art, sculpture and materials.

'I followed this up with a three-year Diploma in Art and Design which placed emphasis on manual skills especially in metal and silverwork.

'I found this a bit limiting so I decided to broaden my experience by enrolling for a sandwich course in industrial product design in Birmingham. This included two industrial placements: six months with a cutlery manufacturer in Birmingham and another six months with a manufacturer in southern Germany specialising in glassware, cutlery and commercial catering equipment.

'It was interesting to compare the practices in both companies, and by the end I was beginning to understand industrial design and how to work out the correct qualities of materials for producing goods.

'Since the jobs market was looking bleak when I finished my course, I set up my own workshop making pewter tableware. I chose pewter because it is an easy material to work with and requires simple tools and techniques. I also did some part-time lecturing, which led me eventually to a full-time lectureship in art and design at Hong Kong Polytechnic.

'I returned to Britain and joined the Design Council in London. Here I worked as an industrial officer for the ceramic, glassware and metal industries. My remit was to work with companies and encourage them to be more innovative and design conscious. It has to be said that not many were in the habit of using designers in those days.

'After eight years with the Design Council, as I had no wish to move up into management, I decided to set up on my own. I became a design consultant for two design companies. When work tailed off in the early 1990s I started to design and make luxury pens and pencils for sale. Other kinds of giftware for men followed.

'To be a good industrial designer you have to have an understanding not only of graphic design (two-dimensional design) but also shapes, form, scale and proportion (three-dimensional design). You also need to know how things are made and which materials are best suited for a particular job.

'I normally design the prototype of a product and then go to talk to producers I know. One firm, for instance, turns the metal components for the pens and pencils, another does the plating and hand polishing, and I use a leather company to put leather covers on some of the pens. Some of the assembly work I do myself; the rest is subcontracted.

'I began to acquire customers by exhibiting at trade shows in the UK. Later I started attending similar shows in Germany with a colleague and found the Germans very interested in my products. Soon I was exporting my products – not only to Germany, but also the USA and Japan where I have my own website in Japanese (www.davidhaywarddesign.jp) in addition to my UK one (www.davidhayward.com).

'I moved from my small workshop into a studio shop, and decided to branch out into jewellery, which I currently only sell through the shop.

'I tend to be pessimistic about some of the manufacturing sectors in the UK. However, the outlook for small-scale producers is much more positive. A lot of designers are beginning to make their own products, just as I have done, and the number of designer-craftsmen is increasing and flourishing.'

CHAPTER 2
WHAT'S THE STORY?

Manufacturing used to involve making things by hand. People worked in workshops stitching clothes, making shoes on a last, making pottery on a potter's wheel and firing it in a kiln, making chairs and tables, and so on.

With the Industrial Revolution in the eighteenth and nineteenth centuries machinery was introduced into manufacturing, which meant goods could be produced more quickly and cheaply. Large swathes of manufacturing moved out of small workshops into large factories and the era of mass production was born.

These days we tend to think of manufacturing in terms of large factories with long assembly lines manned by operatives who check the machines periodically to make sure they are functioning correctly.

However that is not the whole story. Manufacturers come in all shapes and sizes and the majority of them are very small. In the West Midlands, for instance, there are thousands of small engineering companies making components for the car industry. Sixty per cent of firms making furniture have fewer than nine employees and in the printing industry 90% have fewer than 20 employees.

Larger firms often manufacture a range of different products and have several layers of management. They may well have their own in-house sales and marketing departments as well as their own product development sections which involve designing and making improvements to products and then testing them to see if they are fit for purpose.

Where the manufacturing process is highly mechanised plant maintenance engineers play a crucial role, and there are people involved in specialist roles, such as production schedulers, purchasing officers, logistics personnel and personnel officers.

Britain makes virtually everything – from trumpets to toy soldiers – and it is impossible to estimate how many different product lines come out of the country's factories and workshops. This chapter aims to give you an overview of the principal branches of manufacturing and the variety of jobs they offer.

QUIZ

Here are the names of some leading manufacturers. What do they make?

1. Astra Zeneca
2. BAE Systems
3. Cadbury
4. Conway Stewart
5. Coats
6. Corus
7. Dunlop
8. GKN
9. Norcros
10. Pilkington
11. Rolls-Royce
12. Unilever

ANSWERS

1. Pharmaceuticals
2. Aerospace and defence equipment
3. Chocolate and confectionery
4. Pens
5. Textiles (including thread and clothing)
6. Steel and steel products
7. Tyres, sports equipment
8. Car and vehicle components
9. Ceramic products, including tiles
10. Glass
11. Aero-engines, marine engines, etc
12. Food and household products

All of the following career sectors are covered in the 14–19 Diploma in Manufacturing and Product Design.

FOOD AND DRINK

This sector accounts for around 15% of manufacturing, employs over 390,000 people and has a turnover of £77.4 billion. It is a vast industry with a number of large manufacturers, such as Unilever (think of Birds Eye and Walls), Northern Foods, Cadbury, United Biscuits and plenty of much smaller ones (such as craft bakeries and microbreweries).

Some manufacturers specialise in one aspect of food production, such as baking (biscuits, bread, etc), freezing (peas, fish, etc),

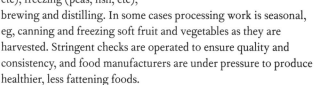

DID YOU KNOW?

20% of UK expenditure goes on food – £150 billion. We spend nearly £4.5 billion on chocolate and another £9 billion on soft drinks.

brewing and distilling. In some cases processing work is seasonal, eg, canning and freezing soft fruit and vegetables as they are harvested. Stringent checks are operated to ensure quality and consistency, and food manufacturers are under pressure to produce healthier, less fattening foods.

British food and drink products are exported all over the world. Scotch whisky is a particularly successful export and new distilleries are having to open to cope with demand from consumers in the Far East.

TEXTILES, CLOTHING, FOOTWEAR

This sector, which includes leather products, accounts for 3.3% of UK manufacturing output, and employs nearly 100,000 people with a range of skills. Making clothes is a succession of processes from spinning the yarn, weaving it into cloth, cutting the cloth into shapes and stitching these shapes together to make the finished product. Knitted products undergo a similar process. Dyeing and printing are other key processes.

Design is important at the top end of the market, and London has become one of the world's centres of fashion design. The textile industry faces stiff competition from clothing manufacturers in Asia and other countries where labour costs are lower. Even so sales exceed £9 billion and exports hover around the £6 billion mark. Nowadays, Britain specialises in up-market, high-quality fabrics and clothes.

The industry is centred mainly in Yorkshire and Lancashire as well as the north of Scotland. Bespoke tailoring and dressmaking is found all over the country, though design tends to gravitate towards the capital. Footwear manufacture has declined in the face of cheap imports from abroad, but still employs around 3,500. It is based predominantly in Northamptonshire and concentrates on the quality end of the market.

FURNITURE

The furniture industry employs over 90,000 people in around 20,000 companies spread all over the UK, but mostly in London, the South East and North West.

There are many different sectors in this industry which require varied skills and training, however much of the training is done on the job itself. Workers in the industry tend to be specialists in a certain area and work specifically on one type of manufacturing. The three broad areas of furniture are: domestic furniture (beds, chairs and tables, wardrobes and other home furniture), office furniture (including desks, chairs and storage furniture) and contact furniture (which covers hotels, airports, surgeries and other public areas). Within these sectors, craftsmen will specialise in five main areas, including restoration and reproduction, soft furnishings, bathroom fittings, cabinet work (chairs and tables) and upholstery.

DID YOU KNOW?

One of the most famous British furniture makers was Thomas Chippendale, who in the eighteenth century was the first to issue a catalogue showing his designs. His furniture is much prized today.

The industry is changing continually and many jobs have been taken over by machinery, which in turn are managed and programmed by skilled furniture manufacturers. This is true of the mass-manufacturing side of the sector, however self-employed craftsmen continue to manufacture products themselves as the emphasis is on the quality of a certain piece of furniture rather than the quantity produced.

PAPER, PRINTING AND PUBLISHING

With a turnover of £15.2 billion this is the UK's fourth largest manufacturing sector.

The manufacture of pulp, paper and paper products (including cardboard, wallpaper, cartons and tissue paper) employs over 60,000 people mainly in North West England, Yorkshire, East Midlands and the South East.

Printing and publishing accounts for a further 280,000 jobs in many parts of the country. The majority of firms employ under 10 people. The print process needs people with knowledge of design, inks, paper, and the printing presses themselves, as well as people to oversee the whole process, check quality, deal with customers.

DID YOU KNOW?

100 tonnes of ink are used each day in the UK for printing newspapers.

Despite the advent of the Internet and electronic books, the printed word continues to hold its own. Printing technology has undergone several revolutions over the past 50 years and hot metal printing presses are now virtually obsolete. The industry has also become more flexible with the introduction of print on demand technology in certain areas of book publishing.

CHEMICALS AND PHARMACEUTICALS

This is one of Britain's largest manufacturing industries. It has a £50 billion turnover and employs 175,000 people. It is the country's leading exporter and offers pay rates that are on average higher than in most other manufacturing industries to reflect the high skills and productivity of the workforce.

The heaviest concentrations of industry are in Scotland and the north of England and employers range from multinational chemical and oil companies to specialist small and medium-sized businesses.

The pharmaceutical industry alone employs 65,000 people, one-quarter of them graduates. It makes the third highest contribution to Britain's trade balance with exports which exceed £2 billion. Pharmaceutical firms are found in many parts of the UK.

Job prospects in both industries are good and currently there is a shortage of people qualified to S/NVQ level 2 and 3 in the chemical industry compared to the proportion of jobs at those levels. There is a vital need to increase the supply of young people with the right skills and aptitude for the industry.

DID YOU KNOW?

The modern chemical industry in the UK came about in the mid-eighteenth century because of the need to develop rapid bleaching techniques for the cotton industry.

POLYMERS

Polymers are synthetic substances like plastic, nylon and PVC. Plastic has a huge range of uses from carrier bags to electrical cables, and the polymer industry is one of the most important areas of manufacturing to be in. Many household items, from bags to CDs, are plastic. Cars, aeroplanes and even space shuttles all use advanced plastic and polymer composites. As well as processing plastics, the industry also incorporates: rubber manufacture, which is then used in a wide variety of ways from tyre manufacturing to seals, mats and hose pipes, and polymer composites processing, which involves combining and reinforcing plastics with other materials to make a stronger, more durable product. This sector of the industry is expanding rapidly and revolutionising the medical and sports industries. Finally there is the signwriting industry, which is responsible for the production of signs in all different materials for companies, businesses and government organisations.

Over 14,000 companies operate within the sector employing some 170,000 people. The industry is extremely dynamic, being subject to many changes and opportunities arising from technological change, development of new materials and processing technology. Over 1,000 employees are employed in research and development activities.

CERAMICS, GLASS AND NON-METALLIC BUILDING PRODUCTS

Around 95,000 people are employed in this sector, which includes ceramic products, glass and concrete.

Ceramics

Clay is an important raw material which has a variety of uses in the building industry and for domestic use. Different types of clay are used according to the qualities needed in the item to be made, such as strength or delicacy. Ceramic products are used in construction and building, in industry and in the home. The ceramics industry is split into domestic and industrial products. Industrial products include pipes, drains, chimney pots, roofing tiles and components for use in other industries such as agriculture and aerospace products. Many household items are made from ceramic also and these include tableware (cups, plates and bowls), ornamental objects, gardening products and bathroom fittings.

Staffordshire is an important centre for ceramics and many of the large manufacturers are based here, but you will find smaller craft potteries up and down the country. Brickmaking is traditionally associated with Bedfordshire but is also carried out elsewhere. Demand for building products is closely linked to the state of the construction industry.

Glass making

Glass is a versatile material made from silicates and has a variety of applications ranging from windows and glass doors to fibre optic cables, telescopes and microscopes, glassware and glass ornaments. Apart from large manufacturers, such as Pilkington, which specialise in flat glass, there are around 10,000 companies – large and small – making glass and glass products.

Flat glass is used extensively in the construction industry – in doors, windows and conservatories, for instance. The automotive industry is another significant user of glass for windscreens, windows, lights and mirrors.

Many of the goods we buy at the supermarket are packed in glass jars or bottles – and we use glassware extensively in the home and catering establishments and purely for ornamentation.

DID YOU KNOW?

The oldest glass container we know of dates from 1500BC. Glass making in the UK dates back to AD680 in the area around Jarrow and Wearmouth.

There are a number of specialist firms which specialise in making glass products for use in science laboratories, and a glass product, fibreglass, is increasingly used in the making of sailing boats and telecommunications.

METALS AND METAL PRODUCTS

This sector has an annual turnover of £38 billion and employs 450,000 people. They work in companies ranging in size from the giant steelmaking firm of Corus, which has well over 20,000 employees, to small foundries which make metal castings for engineering firms from steel or non-ferrous metals.

It encompasses a wide range of operations from the actual production of the metal to the final finished product. They include pressing and flattening hot metal sheets, pouring molten metal into moulds and cutting metal ingots into shape – often with computer-controlled machinery. After use most metal objects can be melted down again and recycled.

South Wales, North Wales, South Yorkshire and North Lincolnshire are important steel-producing centres. Foundries are found mainly in the West Midlands, North and Central Scotland, but engineering firms making metal components can be found all over the country. The industry's fortunes are closely allied with those of the construction industry and manufacturing.

See Job profiles on Foundry Worker and Sheet Metal Worker on page 27.

ELECTRONICS

Britain's electronics industry is the fifth largest in the world. It is a dynamic and expanding sector of the economy and employs 300,000 in many parts of the country, including the M4 corridor west of London and Silicon Glen in Scotland. It encompasses a wide range of products, such as electric motors, generators and transformers, electricity distribution and control apparatus, insulated wire and cable, office machinery, including photocopiers, cash registers and computer equipment, television and radio receivers, sound or video recording equipment, semi-conductors, communications technology, electronic instrumentation and control equipment.

The majority of employers are small, with 91% of all UK sites employing fewer than 50 people. The UK is a major centre for electronics development companies and home to 40% of semi-conductor design houses.

DID YOU KNOW?

Faraday, Ohm, Volt and Watt have one thing in common. They are all measurements of electricity.

There is a strong requirement for skilled and flexible people in most areas as the industry moves to the manufacture of more high-value products.

AEROSPACE, AUTOMOTIVE, MARINE AND GENERAL MANUFACTURING

Britain is one of the world leaders in **aerospace**, a sector which includes both civil and military aircraft, weapons systems and satellites. The majority of firms are involved in producing components rather than building complete aircraft. They tend to be concentrated in the South East, South West and East of England. The sector employs

DID YOU KNOW?

It takes two tonnes of paint to paint a jumbo jet.

nearly 100,000 people. The distinction between manufacturing and servicing aircraft is becoming blurred.

The **automotive** sector covers the manufacture of motor vehicle bodies, engines, components and accessories, trailers and semi-trailers and employs 150,000 people. The majority of employers within the industry are small, with 82% of all sites employing fewer than 50 people. Most supply components to vehicle manufacturers rather than manufacture vehicles themselves. Although many of the famous British car marques no longer exist, the UK can boast a 30% share of European internal combustion engine production and is particularly strong in motorsport and automotive design engineering. Three-quarters of UK car production is exported.

Q DID YOU KNOW?

Before the nineteenth century ships were built of wood. The first transatlantic liner to be built of iron was Isambard Kingdom Brunel's *SS Great Britain* in 1845, which is now a popular tourist attraction in the port of Bristol, where it was constructed.

The **marine** sector encompasses shipbuilding (including submarines), boatbuilding and marine equipment (including offshore oil platforms). If servicing and repair activities are included, over 55,000 are directly employed in this sector. Yacht building, in particular, is thriving despite strong competition from abroad. Some UK powerboat builders export more than 90% of their production.

CHAPTER 3
WHAT ARE THE JOBS?

Below is an overview of some of the jobs available in the manufacturing and product design industries. It is by no means an exhaustive list, but gives you a good idea of some of the job areas and the type of work involved.

FOOD AND DRINK

Food process worker

The work of a food process worker depends very much on the size of the operation. Large organisations tend to make extensive use of machinery and you could find yourself watching the machines which process the ingredients, adjusting them as and when necessary, ensuring that sufficient raw materials or ingredients are being fed into the process and checking the product at each of its different stages. On the whole this tends to be repetitive work, but you have to know what to do if anything goes wrong and respond accordingly. This might involve stopping the machine and informing your supervisor or the maintenance engineer on duty. Everyone is responsible for keeping the machines and working area clean.

In smaller outfits many more of the operations are carried out by hand, but there is usually some machinery involved in the process. The work could involve weighing out ingredients, mixing them, checking the temperature gauges where cooking is involved, and packing the finished product.

In order to get a job in a food processing operation, it would be helpful to have qualifications in English, maths, technology and food technology, especially if you are hoping for an apprenticeship or eventually a supervisory position. However, not all employers insist on qualifications, as on-the-job training is usually available, which would cover such matters as hygiene, health and safety as well as the manufacturing process.

Brewery worker

If you opt to work in a brewery, particularly one of the smaller, traditional breweries, you are likely to be involved in a variety of different tasks from unloading a delivery of hops and barley, the basic ingredients of beer, to despatching – or even delivering – the finished product.

The first stage involves adding water to the barley and boiling it in a kettle (or copper) and hops are added. The mixture is then transferred to a large vessel known as a fermenter where yeast is added. Fermentation starts converting the mash into alcohol and carbon dioxide. When the brew is ready it is poured into kegs or casks or put into bottles and cans. The bottled or canned beer is then heated to 60° Celsius in order to pasteurise it. The end product is then loaded onto a brewer's dray for delivery to customers.

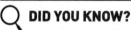

DID YOU KNOW?

Lager is fermented at a lower temperature than beer and the process takes longer.

Large, modern breweries tend to be extensively mechanised with computers which monitor the progress of the beer. Traditional ones require more human intervention, including weighing out the ingredients, monitoring the progress of the beer, operating machinery, filling the barrels and loading them onto lorries, which can be physically demanding.

Scrupulous attention to hygiene and accuracy is required of anyone involved in the actual brewing process, otherwise the beer could go off. Employees also need to be extremely safety conscious.

TEXTILES, CLOTHING AND FOOTWEAR

Textile operative

As you will have seen in Chapter 2 the textile industry encompasses a wide range of products besides the obvious ones, such as cotton thread, knitting wool and fabrics which are turned into clothes, soft furnishings and upholstery. Textiles are also used in construction (eg, roofing felt), hospitals (bandages and other dressings) and even in aircraft manufacture.

There are generally reckoned to be four stages in the production process: the preparation of fibres; spinning fibres; fabric production; and the finishing process. It would be unusual for all stages to be carried out on the same premises.

The first stage of the process involves dealing with the raw material, cleaning, combing and twisting the fibres into yarns – called slivers in the trade. The slivers are then spun and wound onto bobbins, and often dyed during this process. Next the yarn is either woven or knitted into a fabric. Alternatively it is felted or bonded into a non-woven fabric, such as twill or satin. The finishing process might involve further treatments to the fabric – such as printing, fireproofing or dyeing.

These processes are highly automated and as an operative you could be overseeing a number of machines, making adjustments where necessary and keeping a record of what has been produced or processed.

Sewing machinist

Making clothes, often described as apparel in the trade, often involves taking the fabric, marking it out and then cutting the fabric into the marked sections. The pieces are then passed to the sewing machinist who turns the cloth into garments, after which they are pressed, packed and despatched.

Some sewing machinists make up garments themselves using industrial sewing machines to stitch the pieces together, make buttonholes and perform other tasks. Increasingly, however, such work is carried out by computer-controlled machines. These need to be monitored and adjusted and checks have to be made that the finished product matches the original pattern.

Skilled machinists, known as sample machinists, are closely involved in the design process. They make the prototypes of garments and other products to assess whether or not they are viable.

Shoemaker

Most of British footwear production requires people with craft skills who can use hand tools as well as semi-automated equipment, such as cutting, sewing and polishing machines. High standards are required as British shoe production is for the more expensive end of the market.

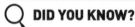

DID YOU KNOW?

Much of the 2005 film *Kinky Boots*, which starred Chiwetel Ejiofor, was shot on location at a shoe factory in Earls Barton, Northamptonshire.

There are several stages in shoe manufacture which are either performed completely by one craftsperson or split up between two or more.

First, the leather or other material for the uppers has to be trimmed – a process known in the trade as 'clicking'. After that the leather is sewn and treated. This is known as 'closing'. Next the uppers are shaped ('lasted') and the soles, insoles and heels are added – either by a welding process or by further stitching (a process known as 'making'). Lastly the shoes are 'finished': they are trimmed and polished, and laces are added.

FURNITURE

Cabinet maker

In modern furniture factories employees tend to assemble furniture from ready-made components, checking to ensure that every piece is a good fit and taking corrective action if it is not. The job may entail using hand tools or mechanical tools, glue, screws or other kinds of fittings. Manual dexterity is a must.

Cabinet makers tend to be at the luxury end of the furniture market and are normally closely involved in the design of the furniture they make. The work will normally consist of making a design in consultation with the client, estimating the amount of material to be used, selecting the material (usually wood), and offering a quotation.

The next stage is to order the wood, cut and shape it to the stated requirements, and then fix the parts together, adding hinges, handles and locks where necessary. Polishing the finished article may be done in house or contracted out to a specialist. Highly skilled cabinet makers are also employed in furniture restoration.

Upholsterer

An upholsterer takes the basic frame of a piece of furniture, such as a sofa or chair, pads it out, fits webbing and springs to it, and covers it with some form of material ranging from a textile fabric to plastic or leather.

If you work in a furniture factory many of the decisions on patterns, fabric, cushions and trimmings will be made for you, and you will have to be able to read instructions and design plans. Accuracy and attention to detail are important.

Many upholsterers work in small workshops making furniture to order, and may find themselves advising clients on furniture design and preparing estimates on the cost of the job. The main tasks are cutting the material, stretching it over the frame, stitching or stapling it to the frame, adding trimmings, and perhaps adding castors to the legs.

Upholsterers do not only manufacture new furniture. Some workshops specialise in renovating and restoring old furniture, some of which may be very valuable.

CHEMICALS AND PHARMACEUTICALS

Chemical plant process worker

The actual work of a process worker will differ markedly according to the size of the operation and the end product. A large chemical plant which operates on a vast site will normally have a central operations room with computer screens and gauges that provide information on how the different sections of the plant function. Staff need to monitor this information, often on a 24-hour basis, maintain a log book, make adjustments to the functioning of the equipment, report faults, and react to any emergencies.

In smaller factories process workers operate individual machines, ensuring the correct amounts and proportions of raw material are fed into the process and that the machinery is functioning efficiently. You need to be alert to possible problems and if you spot a malfunction either fix the problem yourself or report it to the maintenance engineer. During the process samples of the product are often taken and tested to ensure that the quality is acceptable.

Process workers often have to ensure the cleanliness of the equipment used and prepare machines for each new batch. In the case of smaller items they may also have to supervise the packaging process.

Laboratory technician

Laboratory technicians play a crucial role in many areas of manufacturing. In the laboratories of pharmaceutical firms, for instance, they test and check products and support the development of new products.

The job could well involve maintaining and cleaning equipment, checking and reordering stock, setting up experiments, recording data and products testing. Accuracy and attention to detail are essential.

Many organisations are prepared to accept people with GCSEs and offer them on-the-job training, but increasingly manufacturers prefer to recruit candidates with higher-level qualifications.

CERAMICS, GLASS AND NON-METALLIC BUILDING PRODUCTS

Ceramic worker/potter

As we saw in the previous chapter, this industry embraces a wide range of products – from pots to paving stones. Pottery workers are employed in a variety of environments from large highly automated factories to small workshops staffed by just one or two people. In the latter arrangement a person may be involved in all aspects of crafting the product from the design stage to the baking process (or 'firing') in a hot kiln to harden it.

In a highly mechanised industrial situation you could be controlling machines which pour the liquid clay (known in the business as 'slip') into moulds. This process is known as 'casting'. After the product has been shaped it may be coated with a glaze to make it waterproof, as in the case of sanitary ware. Decoration is important for tableware (cups, saucers and plates) and transfers are applied to the product at this stage. High-value tableware can be handpainted.

The final stage involves placing the product in a kiln and monitoring the firing process. Products have to be carefully checked at the end of the process and imperfect ones discarded.

In traditional potteries a number of different techniques are employed. Making flat objects on a rotating mould is called 'jiggering'; shaping clay on a traditional potter's wheel is known as 'throwing'; inserting a metal tool into a mass of clay to hollow it out is called 'jolleying'.

Glass maker

Glass makers work in a variety of environments. You could find yourself working in a large factory using computer-controlled machinery making glass containers by the million or flat glass. Alternatively, you might be at the craft end of the business designing and making glass ornaments, stained glass or scientific instruments in a small workshop.

The process starts by heating silica, soda and lime and other additives with scrap glass (known as 'cullet'). To make flat glass the preferred method is to pour a stream of molten glass through rollers, after which it is floated over a bath of molten tin. To make bottles a stream of molten glass is cut into 'gobs', which are placed in a mould and have compressed air blown into them. For glass containers the molten iron is poured into cast iron moulds. All these processes are highly automated and people operating these machines have

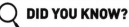
DID YOU KNOW?

Two-thirds of all glass produced in the UK is used to make containers – jars and bottles – some 2 million tonnes in all?

to be scrupulously observant and be prepared to react quickly if problems occur. Once ready the glass may need to be cut, engraved, ground into shape and subjected to a number of other processes.

A distinction is drawn between glass manufacturing and glass processing. The latter (which includes glass blowing, scientific glass making and glass decorating) requires skills of a high order. The main centres for glass making are Yorkshire, the West Midlands, London and parts of Scotland.

PAPER, PRINTING AND PUBLISHING

Paper manufacturing operative

Most paper mills are wholly automated these days and a typical operative's day would involve monitoring the progress of the paper as it passes through machines and taking appropriate action in the event of a mechanical breakdown.

The first process involves feeding the raw material into the paper processing machine. This will normally be wood pulp or paper for recycling. In the case of recycling paper care has to taken to separate out extraneous materials, such as plastics, which could contaminate the pulp or cause mechanical problems.

The ingredients are then mixed with water, caustic soda and phosphates and heated to a temperature of between 65° and 90° Celsius. This is known as the 'wet end' of paper manufacture. As an operative you will need to check that the pulp mixture attains the right consistency before its passes on to the next stage.

There is often a separation of duties between the 'wet end' and the 'dry end' of paper manufacturing, when the pulp is fed into the paper-making machines. Here the paper is normally passed through rollers which press out the water. A further process involves drying the paper out by means of evaporation and 'sizing' it (to make it non-absorbent) before winding it on to rolls. Again, the machinery needs to be closely watched to check the quality of the paper and ensure that no tearing occurs as it is fed on to rolls.

After a final check to ensure the product meets the customers' requirements it is either placed in storage or prepared for dispatch.

SIGNMAKING

Signwriter/signmaker

Signwriters and signmakers make a considerable contribution to our visual environment. We see their handiwork everywhere, both indoors and out of doors – from road signs to shopfronts, from pub signs to banners. The signs we see around are sourced from a wide range of materials such as vinyl, PVC, steel, aluminium and wood.

Some signwriters are involved in the design of signs right from the beginning, but in other cases they work from plans submitted to them by clients or designers. Traditional ones may paint directly onto the material they are using while others may use stencils for the lettering and other aspects of the design.

Increasingly computers are being used in signmaking; and in the case of road signs, for instance, the words are printed onto the metal background using lasers. Carpentry and metal cutting skills are employed for cutting signs to the required size and making letters for three-dimensional installations.

Signmakers may be required to install or supervise the installation of their signs, which can involve physical exertion and working at heights, but this is not always the case. They may also need to consult with clients and check that the signs they put up do not breach any regulations.

METALS AND METAL PRODUCTS

Foundry worker

In this process metal ingots are melted down and the molten metal is poured into moulds or hollowed out shapes made of sand to make components primarily for the engineering industry. However, there are also specialist foundries which turn out church bells, metal sculptures, etc.

Foundry workers can be engaged in a variety of tasks, heating the metal, preparing the moulds, and trimming off the finished product as it cools down. Among the varied production techniques are greensand moulding (which involves preparing a mould from sand), die casting, shell moulding, investment casting and lost wax casting.

Although traditional foundries still exist where the casting is done mainly by hand and considerable teamwork is required together with plenty of lifting, machinery has revolutionised the industry. Today's foundry worker is more likely to be operating computer-controlled equipment and ensuring that the right temperatures are maintained.

Sheet metal worker

Sheet metal workers work on the metal sheets that arrive from the steel mills and transform them into useable products, such as car panels for the automotive industry or steel girders for the construction industry.

The work typically involves marking out and then cutting sections to the required size, shaping them, finishing them and perhaps assembling them. Increasingly computerised machines are used to do this work, leaving more specialist matters to craftspeople using hand tools to trim off rough edges and ensure that sections have been bolted or riveted together correctly.

In heavy industry, such as ship building, sheet metal workers are known as platers. Their job is to weld or rivet thick metal plates to build ships' hulls and drilling rig platforms. This kind of work is found chiefly in Scotland and in the north of England.

Sheet metal workers and platers sometimes have to work shifts. The work can be physically demanding and take place in a noisy environment, both indoors and out of doors. Most sheet metal work jobs are in the Midlands, North East, South Wales, London and the South East.

AEROSPACE, AUTOMOTIVE, MARINE AND GENERAL MANUFACTURING

Aerospace engineering technician

This is a highly responsible job which could involve designing, building and testing new aircraft and components. Servicing and repairing may well form part of the job.

A technician would normally work in one of two areas: mechanics – building and servicing aircraft fuselages, or hydraulic and pneumatic systems, such as wings, engines and landing gear or avionics – installing and testing electrical and electronic systems used in navigation, communications and flight control. Workers in this sector will have a good working knowledge of CAD (computer-aided design) and CAM (computer-aided manufacture) as much of their time will be taken up with using these methods to develop components.

DID YOU KNOW?
Thirty per cent of employees in the aerospace industry are educated to degree standard or the equivalent.

The jobs in both sectors also involve investigating and testing solutions to engineering problems caused by weight, altitude, temperature and engine performance as well as building and testing prototypes using computer simulations and physical models and using prototypes to predict and refine the performance of aircraft systems.

Engineering craft machinist

Whereas some engineering components are cast from moulds, others have to be cut out of the raw materials (metals, plastics, etc) and shaped into the required form using lathes, presses, drills, milling machines, grinding machines, cutters and other precision instruments.

Car assembly worker

The motor industry and various other engineering manufacturing organisations employ large numbers of people to assemble hundreds of components into a finished product.

In large factories workers perform a particular task or a number of predetermined tasks in the assembly of a product - usually on an assembly line. The work may involve working with one's hands only or using different hand tools such as drills, pliers, screwdrivers, soldering irons and special tools to position and hold the parts in place during assembly. Much of the work is performed standing, and assemblers may be rotated so that their work does not become too repetitive.

You will find a number of other specialist jobs in car factories, such as toolsetters (who change the tools and settings on computerised machines), technicians and engineers who ensure that the machinery is working efficiently. Many factories operate shift systems.

Some of these tasks are performed using handtools, especially if only small batches are required. But increasingly computer numerically controlled (CNC) machines are used which need to be carefully adjusted at the start of each batch. But before this stage is reached the craftsman has to interpret the technical drawings and instructions which have been supplied and convert them into a computer program which is then fed into the machine.

The machining process has to be monitored carefully and adjustments made as and when necessary. Samples have to be taken of the end products, which are checked carefully against the original specification and either adjusted or rejected if they do not. See the case study in Chapter 5 for a more detailed look at this sector.

HIGH-LEVEL SPECIALIST JOBS IN MANUFACTURING

Product designer

Product designers combine technical knowledge with design skills in order to make improvements to products currently on the market or to design brand new ones. Some may work directly for a manufacturing company or alternatively they may work for a design studio which has a design contract with a manufacturer.

There are a number of stages in industrial design. The designer may have an idea for a brilliant new product, but first of all they have to establish that there is a market for it and that there are no competitors in the field to be aware of.

They have to consult widely with potential clients and engineers to test the feasibility of the design brief, liaise with suppliers to find materials which are suitable and cost-effective, produce a design and have a model or prototype made, test the prototype, making modifications if required, and make presentations to interested parties (see Chapter 1, 'Success story').

Production planner

The manufacturing process has to be carefully planned in order
to ensure goods are produced efficiently, cost-effectively, in the
right qualities and on time. This is where the production planner
comes in.

Initially the role involves considerable consultation – with
clients to understand their requirements in terms of product and
delivery times, with section heads and supervisors to find out the
availability of staff and machinery, and with purchasing officers to
establish a delivery time for the raw materials.

On the basis of the information received, a production planner
works out a production schedule for the particular job and
calculates how long the work will take. They then have to check
regularly that the schedule is being maintained and take action to
overcome any problems which threaten to interrupt it.

The position may be office based, but a production planner
will spend time on the shop floor meeting other managers and
supervisors. An understanding of the manufacturing process,
organisational ability and problem-solving skills are vital. A
professional qualification from the Institute of Operations
Management is highly desirable.

Purchasing officer

Manufacturers rely on buying in raw materials and components
in the right quantities and at the right price. It is quite usual to
appoint an individual (or a team) with detailed knowledge of the
business to handle this important function. The remit could
extend to purchasing stationery for the office or supplies for
the staff restaurant.

Officers need to establish the purchasing requirements of the
various departments, consolidate these into orders and then
seek out suppliers who can provide the necessary materials for a

competitive price within an acceptable time frame. In many cases a reordering system has to be implemented to ensure that stock levels are maintained. Accurate record keeping of transactions is essential.

Much of the work is desk based, but a purchasing officer may need to travel to meet suppliers and have discussions with people on the shop floor. The job requires commercial awareness, planning and organisational ability, communication and negotiating skills. Many officers study for the professional qualifications of the Chartered Institute of Purchasing and Supply.

Quality control inspector

In a competitive manufacturing environment quality control is vital to ensure that goods meet specified standards at each stage of the production process and before they are sent out to customers.

The work varies according to the type of industry. Normally products have to be inspected visually to ensure they are the right shape, colour, and/or consistency, and in some cases measuring or testing equipment is used. Quality control inspectors have to keep accurate records and often write detailed reports which analyse shortcomings and propose solutions. In many cases this could involve discussions with supervisors and other staff with a view to improving work practices.

Among the necessary attributes in this job are good vision, an eye for detail, a commitment to quality and problem-solving ability. Many quality control inspectors are members of the Institute of Quality Assurance.

QUIZ

Over the last few pages you've had a lot to take in. Just for fun here's a little quiz which will help you review what you have just learnt.

1 **Where would you find a plater working?**
A. In a pottery factory
B. In a laboratory
C. On an oil rig
D. In food processing

2 **Name a well-known furniture designer and manufacturer.**
A. James Watt
B. Josiah Wedgwood
C. Thomas Chippendale
D. John Constable

3 **In which activity might you use tweezers?**
A. In foundry work
B. In electronics assembly
C. In glass making
D. In boatbuilding

4 **Where is die-casting performed?**
A. In the textile industry
B. In the chemical industry
C. In metal foundries
D. In drinks manufacture

5 **If you overheard people talking about the wet end and the dry end, what would they be referring to?**
A. Paper manufacture
B. Swimming pool manufacture

C. Ship building

D. Glass making

6 What kind of worker would use a last?

A. A food processing operative

B. A car assembly worker

C. A glass blower

D. A shoemaker

7 What is cullet?

A. Scrap glass

B. A food product

C. A special kind of textile

D. A type of polymer

8 What product is clay NOT used in?

A. Washbasins

B. Roofing tiles

C. Windows

D. Bricks

9 Where would you see people jiggering?

A. In a steel making plant

B. In aerospace manufacture

C. In a textile laboratory

D. In pottery making

10 Which of these tools would an engineering machinist be likely to use?

A. A grinding machine

B. A lathe

C. A drill

D. A cutting machine

ANSWERS

1. C
2. C
3. B
4. C
5. A

6. D
7. A
8. C
9. D
10. All of these.

CHAPTER 4
TOOLS OF THE TRADE

You have been reading about a number of different jobs in manufacturing and possibly come across some which would suit you very well. But how well would you fit into your chosen industry? Could you convince an employer that you have what it takes? This chapter may help you to make up your mind.

Look at the following statements and decide which ones best describe you.

1. I'm a very practical sort of person.

2. I want a nine to five job.

3. I'm no good at working under pressure.

4. I like playing about with computers.

5. I have difficulty in concentrating.

6. I like working with other people.

7. I want a job with a future – not a dead end job.

8. I'm hopelessly disorganised.

9. I don't really like change.

10. I like solving problems.

11. I have pretty good eyesight.

12. I can't see any point in gaining qualifications.

Now read the following comments:

1. If you have a practical turn of mind you will enjoy working in a manufacturing environment. This is very much hands-on work. You will be making things and have something to see for your efforts.

2. Such jobs do exist in manufacturing – but these are mainly in offices. Most factories and workshops start their day earlier, though as compensation some shut early on Fridays. Large manufacturing plants often operate 24 hours a day and require people to work shifts.

3. Many manufacturing plants are well organised, but that doesn't mean you can be completely laid back. There are usually deadlines and quotas to be met and when a manufacturer has an urgent order to complete the pressure is on. If you are working on an assembly line you will need to maintain a constant work rate.

4. This is a point in your favour. Computers are increasingly important in manufacturing and you may be using machines with complex computer-controlled systems. Many manufacturing firms, particularly technologically advanced ones, now expect applicants for jobs to be computer literate.

5. You cannot afford to make mistakes in manufacturing, so good powers of concentration are necessary in most of the tasks you will be required to undertake, even in highly automated plants. You need to be able to spot immediately if something is going wrong.

6. Many jobs in manufacturing will require you to work as part of a team and you will need to communicate with different kinds of people (your colleagues, customers, suppliers, for example) in order to get the job done.

7. There are excellent prospects in many manufacturing sectors for people with ability and initiative. Certain areas of manufacturing suffer from a shortage of qualified people, and if you are suitably qualified (or are willing to be trained up) there are good prospects for promotion.

8. You have to be a pretty well-organised person to work in manufacturing. You will need to be punctual, observe safety regulations, follow instructions carefully, and work accurately and consistently in order to ensure that the products you turn out are of a good standard.

9. Manufacturing is a competitive business and firms are forever striving for better efficiency and improved quality. This means they have to embrace new technologies and adopt new operating techniques. This is a sector where you have to be able to adjust and learn new tricks.

10. Problem-solving is the name of the game, particularly if you are involved in research and development. But on the shop floor, too, if a problem arises, you need to be able to analyse what has gone wrong and work out a solution. The ability to think on your feet is a definite bonus.

11. In most sectors of manufacturing good eyesight (including good colour vision) is essential, as you may well be required to examine products for potential faults and reject those which have flaws as part of the quality control process.

12. Qualifications are essential these days and vocational qualifications gained on the job are an indicator that you have expertise in this kind of work. They are widely recognised, so if you decide to change jobs, any prospective employer will ask about them before taking you on.

SKILLS AND ATTRIBUTES

Here are examples of the types of skills and attributes that are required in certain jobs in the manufacturing and product design industries. Go through each of these jobs in turn and see if your own personal traits and abilities would suit the job in question. Of course, many of the more generic skills and attributes are applicable to all the jobs mentioned below.

Food process worker

Workers in this industry have to be very good at working in a team, both small and large. This is because they are required to communicate regularly with each other to ensure everyone is aware of what each other is doing and the production line doesn't break down. Working in a factory also means that you will need to be aware of your own personal safety and that of your fellow workers. This is also true of your personal hygiene; working with food requires a meticulous attitude to this in particular. The job is dependent on people who can stay focused, often under pressure, so you will need to be able to concentrate for a long period of time and maintain a high level of accuracy in your work. You will of course need to be ready to listen and follow both spoken and written instructions but this doesn't mean you can always rely on this; make sure you can take the initiative and prove yourself to be a quick thinker. Although it is not essential, workers in this industry will often have a qualification in food technology and will receive training to gain a hygiene awareness certificate if they haven't already got one.

Cabinet maker

If you are interested in this career you will have already realised that the most important skills you require are solid carpentry and design skills. Added to this, a strong mathematical brain, good attention to detail and the ability to use hi-tech tools and equipment are extremely useful. You need to understand how the materials you are using can be manipulated into the finished product, therefore the ability to work from drawings as well as visualising the final article is fundamental. Often, cabinet makers work for a larger company so you need to communicate regularly with your colleagues to ensure the project you are working on continues smoothly. However, it is also common for cabinet makers to be self-employed so your customer service skills need to be good. As a self-employed cabinet maker your work may change regularly depending on the wants and needs of your customers, so you have to be adaptable and happy to go with what the customer wants even if you do not agree with their tastes or ideas.

Aerospace production worker

Workers in this sector are engineering experts so you will have to demonstrate a broad knowledge of engineering techniques and drawings and preferably have an engineering qualification as well as qualifications in maths, science and IT. On top of this a solid understanding of the technology used in aerospace production is key to being successful in this career. As well as these specific skills, workers in the industry are used to solving problems quickly and effectively and have extremely good practical and methodical working skills. As you will be part of a larger production process you must be able to communicate effectively with your team; this is also vital with regard to your personal safety and that of those around you.

Sheet metal worker

There are a lot of physical demands placed upon people working in this industry. Unfortunately, if you do not consider yourself in good physical shape with normal colour vision and you are not comfortable with heights this may not be the right job for you. Sheet metal workers display very good manual dexterity and must be able to concentrate well and act with extreme precision in their work. As well as being physically demanding, work in this industry also requires the ability to follow working drawings as well as instructions. You are likely to work unsupervised so you must be confident in what you are doing and take control of your own personal safety. Equipment in this sector is very specialised so if you already have experience of working with welding and cutting tools this will greatly help you when you enter the industry and it is also beneficial to possess good IT and numeracy skills.

Engineering machinist

As an engineering machinist, you can expect to need a very logical brain and to be able to work methodically, following instructions and maintaining high attention to detail. It is also very important that you have good hand to eye coordination so that you can work quickly and safely. It is likely that you will already have some knowledge of engineering drawings if you are interested in this job sector and this is very important in order to demonstrate that you can work to someone else's diagrams and specifications. If problems arise you need to communicate them quickly and effectively; remember that you will be part of a larger process and one cog can slow down or stop the whole machine. Make sure you are happy to do everything you can to ensure you don't slow down or stop the process and that you can work independently and keep close contact with the overall production line (if there is one).

Chemical plant process worker

As a chemical plant process worker you can expect to work as part of a team, so you need to be a good communicator and be happy to look out for the safety of those around you. You must be able to work independently as well as in a team so you should demonstrate good initiative and patience. As there can be dangerous elements to the job, you need to be able to react appropriately to any emergency situation should it arise. Other important skills you should possess include good numeracy and literacy skills and a strong grasp of IT.

Signwriter/signmaker

Signwriters are creative and inventive with a good eye for detail but it is also important to have sound business skills and a practical mentality. Accurate spelling and good numeracy skills are integral to this job and you also need to demonstrate an aesthetic sensibility and the appreciation of different fonts, typefaces and materials. Full colour vision is also essential. Signwriters will mostly work to instructions set by someone else or a company, so you must be willing to work to someone else's brief. Signwriting is becoming increasingly IT based so you should have some knowledge of computer design programs.

Of course, you cannot expect to be offered a job on these attributes alone. You will need to undergo training and obtain relevant qualifications if you want to succeed in these jobs.

CHAPTER 5
CASE STUDY 1

LESLIE HUGHES

Production engineer

By no means is all manufacturing done in large factories. Leslie Hughes owns a small precision engineering company which makes a wide range of engineering components. His clients range from power generating companies to gearbox makers, owners of specialist cars and luxury goods manufacturers.

'When I left school I enrolled for a full-time course in general engineering at a local college. However, once on the course I began to have second thoughts, especially as I found the maths component something of a struggle.

'Things changed for the better when the college sent me on a work placement to a newly established precision engineering company. I enjoyed the work, got on well with the boss, and he offered me an apprenticeship with the firm. So I changed from a full-time course to an apprenticeship in production engineering and studied one day a week at the college. I eventually achieved a Higher National Diploma.

'Although I gained most of my practical experience on the job, the tuition at college offered me a theoretical background to the work I was doing. But even more important was that I acquired computer programming experience, which would prove very useful when the firm acquired computer-controlled (CNC) machines.

'These new machines have not taken over completely, however, and we still use machines in the workshop which have to be set manually.

'We tend to manufacture products in small quantities – sometimes just two or three, and at the most 500. For example, we make spare parts for Bugatti cars, wheel hubs for Bentley cars, flywheels for Formula 1 racing cars, gear blanks for a major gearbox manufacturer and burners for electricity generating stations. Sometimes I have to make prototypes of new products, which are then sent for testing at a development laboratory.

'The whole process starts when I receive drawings of the components either by email, fax or through the post. I give the clients a quotation which takes into account the cost of the materials and the time the work will take, and if they agree with my price I then order the materials from a steel supplier.

'Many of the items we make start off as solid bars of metal or blocks which can weigh in excess of 50kg. When goods are only required in small quantities it is usually cheaper to machine off the unwanted metal than to make moulds.

'This is a very competitive business, but it is also very interesting, since you are continually having to update your knowledge and learn new tricks. Of course, it has its frustrations, especially when you have a deadline and the machine you are using breaks down. But I would recommend manufacturing to any young person with a practical outlook who has a genuine interest in finding out how things work. It's definitely much more stimulating than working in a bank!

'What does it take to be successful in production engineering? Firstly, I think you need a good basic knowledge of maths and trigonometry, since it is essential to be able to read technical drawings and work out angles. Secondly, you need to be the sort of person who enjoys taking things apart and putting them together again.

'One thing that worries me is that not enough young people are coming into production engineering these days, and I fear there will soon be a shortage of skilled people in this field. This is a pity, for this can be very interesting and rewarding work. I think anyone with a practical outlook should be encouraged to give it a try.'

Leslie's commitment and can-do attitude has worked very much to his advantage. When the owner of the firm retired, Leslie took over the business and it continues to flourish.

CHAPTER 6
FAQs

Any questions? In this chapter we try to answer some of the more obvious ones.

 Where can I get information specifically about jobs in manufacturing and design?

 If you are still at school or college, the first step is to consult your careers adviser. Alternatively, your local Connexions office may be able to help or its website, www.connexions-direct.com.

You could also have a look at the websites of the Sector Skills Councils, which also offer careers advice. The relevant ones are:

▶ Cogent (Chemical and Pharmaceutical manufacture)

▶ Improve (Food and Drink)

▶ Proskills (Building products, glass, paper, furniture)

▶ SEMTA - the Science, Engineering and Marine Training Authority

▶ Skillfast (Textiles)

▶ Creative and Cultural (Design).

You will find details of these in the 'Further Information' section at the end of this book.

Q **Is there any way that I can find out what these jobs are like?**

A Most schools offer a fortnight's work experience as part of the curriculum. Tell your teachers that you would like to gain work experience in manufacturing or product design, and they will make the necessary arrangements.

Q **What qualifications, if any, do I need to work in manufacturing?**

A That depends very much on the job you are after. Some jobs do not require any particular skills and employers are looking for people who are dependable, adaptable, in good health and with a positive attitude to work, and whom they can train for different jobs.

However, the majority of employers these days require evidence that you have a good basic education – although they may waive such conditions for older people with work experience.

If you live in an area where competition for jobs is keen you will find that candidates with GCSEs and vocational qualifications will stand a better chance of landing a job. Also if you are hoping to get an apprenticeship, you will be expected to have a few good GCSEs at the very least. (See Chapter 7, 'Qualifications and Training'.)

Q **What can I expect to earn?**

A Expect to start off at £10,000–£11,000, though as an apprentice you will probably earn less than this. In certain parts of Britain – notably London and the South East – earnings might well be higher than that to reflect the higher cost of living – perhaps £14,000–£15,000.

Pay levels also vary between different industries. Chemical manufacturing and the automotive industry have traditionally paid better than the textile industry. Once qualified you can hope to be paid more, and with promotion annual pay of £20,000 is a distinct possibility. If you move into a management position or more specialist role it could be even higher (see Chapter 7).

Q **What are working conditions like?**

A This varies considerably depending on the nature of the work. Some factory environments (such as a steel plant or iron foundry) can be noisy and dirty and you may need to wear ear protection. Others are quiet and well ventilated, and are pleasant to work in.

Some jobs, particularly those in large manufacturing plants, may require you to work shifts; others operate only during daylight hours. Often the working day starts earlier than it does for office workers.

A 36–39-hour week is the norm these days. Part-time work is also available and there may be opportunities for overtime, particularly if the firm has an urgent order to be fulfilled. Some manufacturers supplement their normal workforce with agency staff at peak periods.

For most jobs you will be expected to wear some form of protective clothing (which could include gloves, boots, protective glasses, overalls and headgear). You also need to comply with safety regulations, and in the food and drink industry in particular you will be expected to adhere rigorously to hygiene regulations.

Q **Which industries can offer me the best job security?**

A Sadly no job is 100% secure, and when a downturn occurs in the economy employers have to shed staff in order to stay in business. If people stop buying cars, car production has to fall and this means redundancies among car workers and the steel workers who provide the automotive industry with its raw materials. Likewise if there is a slump in building jobs, the building products industry may be affected.

Some manufacturing industries are less prone to economic fluctuations. We shall always want food, so food processing is a safer bet than most. We shall always need to wash our hair and clothes, so jobs in the household products sector (which makes shampoos and soap detergents) are also relatively immune to economic fluctuations.

However, the buck stops with you. If you become a key worker in a particular industry and times get tough, your job will be among the last to go.

Isn't working in manufacturing for men only?

 Not at all. Although men outnumber women in manufacturing by a ratio of four to one with regard to full-time jobs, in some industries (eg, printing and publishing) it is more like three to one. Just over one-third of women in manufacturing opt for part-time work.

In any case, the traditional demarcation between men's jobs and women's jobs is less pronounced these days, especially since in many industries machines have taken over much of the heavy work. Employers are much more interested in a person's abilities than his or her gender.

 I don't want to get stuck in the same job all my life. Will I be able to move on into other jobs?

That depends very much on you. If you prove to be a good and reliable employee there may well be opportunities to work in other parts of your company and perhaps overseas, too. In any case, if you are doing an apprenticeship, you will probably be rotated between different jobs and departments as a matter of course in order to give you wide experience.

As you gain in experience and show sufficient aptitude you may get promotion to a more senior position – team leader, for instance.

CHAPTER 7
QUALIFICATIONS AND TRAINING

Manufacturing is a very competitive business these days, and in order to compete globally companies need a skilled workforce. They need people who are flexible, intelligent, able to follow instructions and interpret information. Much of the simple, repetitive, boring work is now performed by computer-controlled machines or robots rather than people.

If you are still at school, you may wonder whether what you are learning has much relevance to working in manufacturing. In fact, employers are looking for people who not only have practical skills, but also good communication skills (which means a good knowledge of English), who can solve problems (including mathematical ones), and who have an understanding of science and technology.

So acquiring a good, rounded education and good qualifications are essential if you are hoping to get a good job in manufacturing industry. That is not to say that personality and motivation are unimportant (they are) – but without the basic skills you might struggle when you start training in earnest for a particular job.

ACADEMIC QUALIFICATIONS

If you are still at school it makes sense to get advice on which subjects would be most beneficial to you. Regard maths, English and a science subject as essential (eg, physics, chemistry, technology). If you are hoping for a job in the chemical or pharmaceutical industries, chemistry would be an excellent choice. If you are interested in food processing, a GCSE in food science or chemistry would be extremely relevant. For some areas of manufacturing a qualification in a craft subject would help you along; woodwork, for instance, would be a useful basis for furniture making. A GCSE in food technology, design and technology or in manufacturing would also get you off to a good start.

Your careers teacher will be in touch with local firms and should be able to advise you which particular subjects they favour when recruiting staff. Another very useful source of information is your local Connexions office, which is accessible via the Connexions Direct website. Don't overlook the websites of the six Sector Skills Councils, which deal with manufacturing and design and the trade organisations of different industries. You will find the addresses of the SSCs and some of the trade bodies in the 'Further Information' section at the end of this book.

 What qualifications, if any, do I need to work in manufacturing?

 That depends very much on the job you are after. Some jobs do not require any particular skills and employers are looking for people who are dependable, adaptable, in good health and with a positive attitude to work, and whom they can train for different jobs.

However, the majority of employers these days require evidence that you have a good basic education – although they may waive such conditions for older people with work experience.

If you live in an area where competition for jobs is keen you will find that candidates with GCSEs and vocational qualifications will stand a better chance of landing a job. Also if you are hoping to get an apprenticeship, you will be expected to have a few good GCSEs at the very least.

Q **What are the most relevant school subjects I can study?**

A GCSEs in English, maths and a science subject are a good start – the more GCSEs you have and the higher the grades you achieve, the better your chances of getting a job. Indeed, if your grades are good enough – A to C in five subjects or more – you should consider applying for an apprenticeship.

In addition to GCSEs there are a number of vocational qualifications you can study for at colleges of further education, such as the City & Guilds Vocational Award in Manufacturing Working Practices and the new Diploma for 14–19-year-olds, which is available at certain schools in England (see page 58).

Q **Some people have told me that NVQs are very important. But what exactly are they?**

A National Vocational Qualifications (in Scotland they are known as SVQs – Scottish Vocational Qualifications) are work-based assessments of a person's skills and abilities. In addition to being assessed trainees have to keep a logbook of the work and training they have undertaken.

There are five NVQ/SVQ levels.

▶ Level 1 indicates you can tackle basic and routine tasks.

▶ Level 2 means you have a broad range of skills and abilities, some of them involving individual responsibility. This is equivalent to five GCSEs at Grade C or above.

▶ Level 3 shows you are a qualified techician and can do complex technical skilled work and supervise others. It is equivalent to two A Levels.

▶ Levels 4 and 5 are degree-level qualifications.

They cover a wide range of subjects. The following list gives you an idea of some of the ones available.

- ► Ceramics Design
- ► Chemical, Pharmaceutical and Petrochemical Operations
- ► Craft Pottery
- ► Electronics Product Assembly
- ► Engineering Production
- ► Food and Drink Manufacturing Operations
- ► Glass Manufacture
- ► Glass Processing
- ► Leather Goods Manufacture
- ► Leather Production
- ► Making and Installing Furniture
- ► Making and Repairing Handcrafted Furniture
- ► Manufacturing Ceramic Products
- ► Manufacturing Sewn Products
- ► Manufacturing Textiles
- ► Metal Processing and Allied Operations
- ► Performing Manufacturing Operations
- ► Process Manufacture (Chemicals)
- ► Processing Operations – Hydrocarbons
- ► Product Development (Apparel)
- ► Sign Making
- ► Textile Manufacturing
- ► Tobacco Processing

VOCATIONAL QUALIFICATIONS

Most schools offer courses in vocational subjects. But when you reach the age of 16 or beyond you may find it preferable to enrol full-time for a vocational course at a college of further education, vocational training centre or the equivalent. The best ones have been designated Centres of Vocational Excellence (CoVEs). Most courses include a work placement so you get hands-on experience as well as a nationally recognised qualification.

The main awarding bodies for vocational subjects are City & Guilds, Edexcel (BTEC) and the Awarding Body Consortium, and their qualifications are recognised all over the country and also abroad. To see what kind of subjects they offer look at Chapter 4, or go to their websites (addresses can be found in the 'Further Information' section).

The training opportunities on offer will vary according to where you live. Colleges tend to offer courses which are relevant to the industries in their catchment area. This means that one of the main centres for courses in ceramics is Stoke on Trent. Similarly, courses in textile manufacturing tend to be concentrated in Lancashire and Yorkshire.

An excellent alternative to a college-based course is an apprenticeship, and these are dealt with later on in this chapter. Younger people from 14 upwards who live in England may wish to investigate a qualification which has been introduced recently: the Diploma.

BTEC qualifications

These are awarded by the examining body Edexcel for the successful completion of courses. There are different levels for these awards.

- ► BTEC Introductory Award/Certificate/Diploma. No formal qualifications are required for entry to this course, which takes you to the equivalent of NVQ Level 1.
- ► BTEC First Diploma takes one year of full-time study, and two years of part-time study and is the equivalent of four GCSEs at Grade C or above (NVQ Level 2).

- ▶ BTEC National Award, which is equivalent of one A Level. You will need four GCSEs grades A–C to start this.
- ▶ BTEC National Certificate. This forms part of a specialist work-related programme and is equivalent to two A Levels.
- ▶ BTEC National Diploma. This is equivalent to three A Levels or an NVQ Level 3.
- ▶ There is also the more advanced Higher National Diploma, which is a degree-level qualification.

BTEC awards are available in a range of subjects, including:

- ▶ Aeronautical Engineering
- ▶ Aerospace Systems Engineering
- ▶ Bakery Technology and Management
- ▶ Biochemical Engineering
- ▶ Ceramic Technology
- ▶ Computer Aided Manufacture
- ▶ Electronic Manufacture Engineering
- ▶ Electronic Publishing
- ▶ Engineering Design
- ▶ Fashion and Textile Design
- ▶ Furniture Design and Construction
- ▶ Industrial Design
- ▶ Manufacturing Technology
- ▶ Marine Technology
- ▶ Metals Technology
- ▶ Printing
- ▶ Printing (ink technology)
- ▶ Print Production
- ▶ Product Design and Manufacture
- ▶ Publication Production and Editing
- ▶ Textiles

Q **Are there any other vocational awards that I could study for?**

 There are plenty. Some are offered by the trade or professional body belonging to a particular industry, such as the Institute of Brewing and Distilling. City & Guilds – apart from offering a wide range of NVQs/SVQs – also offers certificates in subjects such as:

- ▶ Cabinet Making
- ▶ Clothing Machine Mechanics
- ▶ Furniture Production
- ▶ Industrial Ceramics
- ▶ Knitting Machine Mechanics
- ▶ Printing and Graphic Communication
- ▶ Soft Furnishing Skills
- ▶ Staining and French Polishing
- ▶ Textile Techniques
- ▶ Three Dimensional Design and Solid Modelling
- ▶ Three Dimensional Design using AutoCad
- ▶ Two Dimensional Computer Assisted Design

The Awarding Body Consortium offers vocational qualifications across several manufacturing and engineering sectors. Subjects include:

- ▶ Apparel, Footwear and Leather Production
- ▶ Apparel Manufacturing Technology
- ▶ Bespoke Cutting and Tailoring
- ▶ Fabrication and Welding
- ▶ Fashion
- ▶ Industrial Ceramics

▶ Manufacturing Sewn Products

▶ Motor Vehicle Studies

▶ Pattern Cutting

▶ Sewing and Textiles

▶ Textile Technology

The Diploma

The Diploma is aimed at 14–19-year-old students as another option to choose when studying at school and college. So far it is only available in certain schools in England and Wales. It is designed to give students a firmer grasp of the world of work and allow them to train for a specific vocational career, while still having the option to study GCSEs either alongside the diploma or to study the diploma by itself.

The Diploma is available at three different levels for students across the 14–19 age range. Firstly, there is the Foundation Diploma. This is the equivalent to studying five GCSEs grade D–G. The next step up from this is the Higher Diploma, which is the same as studying seven GCSEs and gaining grades ranging from A*–C. The Progression and Advanced Diploma are situated at A Level standard, the first being the equivalent to studying 2.5 A Levels and the latter, 3.5 A Levels. Each level takes two years.

Manufacturing and Product Design is one of a range of Diploma courses and it allows students to develop a variety of skills through hands-on and theoretical learning: from working on a project to carrying out work experience with a company to studying the industry in detail and building a sound knowledge of the career.

If you want to find out more about the subjects available as well as the different learning options for the Diploma go to www.diplomainfo.org.uk.

Apprenticeships

The apprenticeship system is well established and is based on
the principle of earn while you learn. It leads to a vocational
(or work-related) qualification which is recognised nationally.
The apprenticeship system differs according to which country
of the United Kingdom you live in. All areas offer post-16
apprenticeships (which in England are referred to as Advanced
Apprenticeships).

An apprenticeship has three components:

- ▶ Key skills: These are basic skills which are essential if you want
  to be effective in your chosen career. They include numeracy,
  communication, IT, self-improvement in learning and performance,
  and teamwork.
- ▶ A technical certificate: These are usually BTEC or City & Guilds
  awards.
- ▶ A National Vocational Qualification.

In England there are three levels of apprenticeship:

Young Apprenticeships

These are aimed at 14–16-year-olds and offer a work-related
approach to learning. It normally takes two years to complete and
could lead to an apprenticeship or (if you gain particularly good
qualifications) straight on to an Advanced Apprenticeship.

Apprenticeships

These are for people aged 16–24. You develop your skills
within a work environment. Some employers have their own
training centres, on site, but others will send you to an outside
training centre, which may be in a college. It normally takes two
years to complete an apprenticeship and successful completion
enables you to progress to an advanced apprenticeship or direct
into employment.

Advanced Apprenticeships

These are also for people aged 16–24 and are considerably more challenging than an apprenticeship. You may, however, be able to bypass the earlier apprenticeship stages if your qualifications are good enough. It takes between two and four years to complete an Advanced Apprenticeship, at the end of which you will become a fully fledged technician. You could go straight into employment, or if you would like to improve on your qualifications you could consider moving into further education or higher education – perhaps to study for a foundation degree.

Adults without NVQs are also eligible for apprenticeship training under the Government's 'Train to Gain' initiative.

To see how a four-year apprenticeship leading to NVQ Level 3 works, read Daniel Price's experiences in Case Study 3, (Chapter 10).

Apprenticeships in Northern Ireland, Scotland and Wales

In Northern Ireland there are two levels of apprenticeship available for people of all ages.

- ▶ Level 2 leads to an award equivalent to GCSE level.
- ▶ Level 3 leads to an award equivalent to A Level and equates with the Advanced Apprenticeship in England. It is not necessary to have completed Level 2 beforehand if you have suitable qualifications.

Scotland has a scheme called Skillsavers which trains people up to SVQ Level 2 (equivalent to NVQ2). Modern Apprenticeships lead to SVQ Level 3 and are equivalent to the English Advanced Apprenticeship. Your careers teachers or Careers Scotland can advise on the best way forward.

In Wales there is a two-tier apprenticeship scheme.

▶ Level 2 Foundation Modern Apprenticeship leading to NVQ2.

▶ Level 3 Modern Apprenticeship leading to NVQ3.

Apprenticeships: Questions and Answers

What are the age limits for apprenticeships?

The starting age for most apprenticeships was between 16 and 18 in the past, but now there is a drive to update the skills of people of all ages and the age limits have been relaxed.

What qualifications do I need?

To be accepted for an Advanced Apprenticeship (or the equivalent in other parts of the United Kingdom) you should have at least GCSE grade C or above in English, Maths and a science subject. (The Scottish equivalent to GCSE is the Standard Grade.) The more qualifications you have, the better your chances of getting the apprenticeship of your choice. Be warned: competition for apprenticeships is very keen.

What if I don't have these grades?

Some firms will accept you if your grades are lower than this or even if you have no qualifications, provided you pass an aptitude test. (Mature would-be trainees may well be accepted on the strength of their performance in the aptitude test alone.)

How long does it take to complete an apprenticeship?

The completion time for an Advanced Apprenticeship varies according to individual ability. For apprentices aged 16–18 it is normally 3–4 years. Those aged 19 plus may be able to complete their apprenticeship sooner. Other forms of apprenticeship tend to be shorter. However, bear in mind that doing an apprenticeship is not a soft option and there is a considerable drop-out rate.

Q How much time is spent studying at college?

A Normally one day a week will be spent at a college or training centre. (Sometimes attendance at college is done in blocks – a week or month at a time.) The rest of the time is spent getting practical experience on site.

Q What are the normal working hours?

A On days when you are being trained at college or a training centre it will usually be between 9 and 5. If you are working on site you will probably have to start much earlier and the length of the day may vary. Some shiftwork may be involved.

Q Can I get funding for my training?

A In the case of anyone aged 16–18 the cost of the training is fully supported. In the case of older apprentices, the employing company is normally obliged to make a contribution. Adults who lack qualifications are also offered a chance of training for NVQ Level 2 under the Government's 'Train to Gain' initiative.

Q What if I am not accepted for an apprenticeship?

A If the organisation you are working for is unable or unwilling to provide a formal apprenticeship, it might be happy to support you on a relevant training scheme leading to formal qualifications. But don't let rejection put you off. If you don't succeed the first time, try again.

Below is a chart showing the different levels of qualifications

ENTRY/LEVEL 1

Foundation Learning Tier

Level 1

GCSEs grade D–G
Diploma Level 1
NVQ Level 1

Level 2

GCSEs grade A*–C
Diploma Level 2
NVQ Level 2 Apprenticeship

Level 3

General AS and A Levels
Vocational AS and A Levels
Diploma Level 3
Extended Project
NVQ Level 3 Advanced Apprenticeship

Level 4

Degrees
HND Vocational certificates and diplomas
NVQ Level 4

Level 5

Postgraduate certificates
and diplomas, masters
and doctorates
NVQ Level 5

CHAPTER 8
CASE STUDY 2

SUZANNAH HARRISON

Textile industry apprentice

Suzannah lives in Huddersfield and is looking forward to a career in the textile industry.

'In school I became very interested in arts and crafts and did every technology going, including woodwork. The textile industry is very active in my part of Yorkshire, so it seemed the natural choice for me.

'After obtaining GCSEs in English, Maths, Science, German, Graphics and Woodwork I successfully applied for an apprenticeship at the Textile Centre of Excellence in Huddersfield. The centre offers a number of courses, including one in weaving, but I decided to do the Apparel Apprenticeship (Manufacturing Sewn Products option).

'The course had so many different elements and I really enjoyed the variety of activities on offer. At the beginning we were taught how to use printing machines and gained some understanding of dyeing techniques, but most of the time was devoted to pattern cutting and making garments. We learned to make things from all types of materials, including felt and bamboo, using industrial sewing machines.

'Sometimes we did the designs ourselves, but we also worked with different designers. Some of the designing was done on computers (computer-assisted design), and we also did quite a lot of designing using pen and paper.

'The course consisted of a succession of projects in which we learned how to work within guidelines, meet deadlines and prioritise our work. We spent one day each week writing up a diary of the work we had done in order to satisfy NVQ requirements.

'My apprenticeship lasted roughly 18 months and led to NVQ Level 2 in Manufacturing Sewn Products plus the Technical Certificate in Apparel, Leather and Footwear manufacture. Both qualifications were conferred by the Awarding Body Consortium (ABC). I also have a Level 2 award in Health and Safety in the Workplace, which is essential for anyone working in an industrial environment.

'Since finishing my Apparel Apprenticeship I have changed direction slightly to do an Apprenticeship in Business Administration, which includes a City & Guilds NVQ Level 2 in Business Administration, at the Textile Centre of Excellence.

'I'm very keen to continue working in textiles, which – contrary to some people's perceptions – is a very modern and go-ahead industry these days. I feel this additional qualification will give me an even better appreciation of manufacturing industry.'

CHAPTER 9
CAREER OPPORTUNITIES AND PROSPECTS

This book has concentrated on guiding you towards the first rungs of the manufacturing ladder, but the story doesn't end there. However much you like your work, in the fast changing world of manufacturing no job stays the same for the whole of a person's working life. In fact, jobs can disappear completely often as a result of new technology. You don't hear of lamplighters any more, do you?

This means that if you really want to keep on top of the job, sooner or later you will need to undertake further training. This could lead to additional qualifications (an NVQ in another subject perhaps) which would enhance your employability, or you may well wish to go on to study at a higher level.

Another consideration is that your original job could become redundant, either because of reorganisation or because machines or robots are introduced to perform some of the simpler tasks. If this happens, you have to be prepared to retrain.

No occupation is immune to change. Nurses, doctors, scientists, engineers, teachers and electricians all have to update their skills and acquire the latest knowledge from time to time. This process is sometimes known as CPD (continuous professional development).

MOVING SIDEWAYS

Once you have mastered your job you may become aware of other jobs within your particular firm which look more interesting. (The staff notice board is a good source of information on vacancies in other departments.) For instance, you may become interested in product design or the business side.

If you have proved effective in your first job, your company may well look favourably on a request for a transfer as well as offering to equip you with any additional skills you need. It is, after all, in their interest to have a multi-skilled workforce and it could be beneficial for them to transfer someone who knows the company well rather than to recruit an outsider.

On the other hand, if you feel you are getting into a rut, you might consider applying to another company. No two organisations are the same and it could be useful for your future career to gain experience in a number of different environments.

Finding another job should present few difficulties provided you are flexible. Local newspapers, national newspapers, trade journals, government job centres and private recruitment agencies are all worth trying. Also, since this is the age of the internet, be prepared to look at recruitment websites.

Trade associations and trade unions often include job vacancies on their websites, and so do many manufacturing firms. A search engine like Google or Yahoo will lead you to other recruitment websites if you type in 'manufacturing vacancies'. Another good way of finding jobs is to ask around.

MOVING UPWARDS

After you have been with your firm for a while you may have the opportunity to be promoted to a supervisory role, such as team leader, shift supervisor, production supervisor.

Moving to a position of responsibility can be somewhat daunting, and rather than leaving things to chance it makes sense to learn about the art of management. There are a number of useful qualifications that you could study for. Examples are:

- ▶ Introductory Certificate in Team Leading (CMI)
- ▶ Certificate in Team Leading (CMI)
- ▶ NVQ Level 2 Team Leading
- ▶ NVQ Levels 3 and 4 in Team Management.

Many further education colleges organise courses leading to these qualifications, and your employer may be willing to finance the course or even offer you time off in order to study. Extra qualifications like these could well sway the balance in your favour if you apply for promotion or another job.

SPECIALIST JOBS

Your knowledge of how industry works and your experience of dealing with people could prove invaluable if you decide to upgrade your qualifications with a degree or professional qualification. If the safety aspects of the workplace interest you, why not consider becoming a health and safety inspector? If you are a stickler for accuracy and high standards, you might find a great deal of satisfaction if you became a quality control officer.

Instead of assisting designers, you could become a product designer in your own right, but this will involve further study.

If you are an extrovert and enjoy meeting customers, there could be a future for you in sales, and if you enjoy dealing with people, you could consider entering the personnel management field.

Many of our hi-tech manufacturing industries complain of a shortage of qualified engineers, and many are willing to sponsor staff who have proved themselves through an advanced course of study.

CHAPTER 10
CASE STUDY 3

DANIEL PRICE

Apprentice with Unilever

Daniel Price lives in Warrington and is an apprentice with Unilever. In 2008 he was Unilever's Apprentice of the Year.

'During my final year at school Unilever initiated a big drive in the North West to attract apprentices. Since I knew that Unilever has a good reputation for looking after their employees I decided to apply.

'The minimum requirement for an apprenticeship then was a Grade B in GCSE Maths and Grade C in English and a science subject. I had the right qualifications, and they accepted me as an apprentice in the Home and Personal Care Division, which makes soap, shampoos, detergents and other household products.

'For my first year as an apprentice I was sent for full-time training at a small engineering campus attached to West Cheshire College and gained an ONC in Operations Engineering. During my holidays I worked at the Unilever factory in Warrington doing routine maintenance and learning how machines work.

'The second year of my apprenticeship was work-based, but included one week's attendance at college every month. I spent the first six months of the year at Unilever's Warrington plant and was then transferred to the No. 1 Factory in Port Sunlight, which specialises in making liquid products, such as bleach and liquid detergents.

'The factory training was done on a rota system, and I was given the opportunity to work with individuals who know a particular area of operations very well. I was particularly interested to see how the packaging of liquids differed from that of powder products. I also gained an OND in Operations Engineering during the year.

'My third year was split between Port Sunlight No. 3 Factory and Port Sunlight No. 4, where I learned about other manufacturing processes. The first factory compresses powders into tablet form and liquids into gels, while the second specialises in fabric conditioners and washing up liquids. 2008 proved to be a successful year for me and I was nominated Unilever's Apprentice of the Year.

'I am back in Warrington for the fourth and final year of my apprenticeship and studying for an HNC in Operations Engineering and the NVQ Level 3. After that I shall be fully qualified.

'I enjoy working here very much. One reason is that in both Warrington and Port Sunlight you have a closely knit community which is very proud of its industrial heritage. Another is that every day is different.

'I plan to continue working with the company. Given that Unilever has manufacturing operations all over the world, I hope one day I'll be given the chance to work with the company overseas.

'I realise that some areas of manufacturing are having difficulties at present. However, I am confident that there will always be a demand for the kind of products Unilever makes, which means I can look forward to a promising and secure future with the company.'

CHAPTER 11
THE LAST WORD

If you have got to this stage of the book – congratulations!
I hope that you have found something of interest in the preceding
pages – some leads which you think are worth pursuing. If you
want more information, some of the organisations in the 'Further
Information' section of this book will be able to help.

Choosing a job was never an easy task, and many people stumble
upon a career which suits them only after a number of false starts.
By reading this book you will have learned the pros and cons of a
variety of different jobs. Some will sound attractive, but you need
to consider if you have what it takes to make a success of a job in
manufacturing.

Look at the following list. Are you...

Computer literate?	☐ Yes	☐ No
A person with a practical outlook?	☐ Yes	☐ No
Adaptable?	☐ Yes	☐ No
Willing to take on responsibility?	☐ Yes	☐ No
Safety conscious?	☐ Yes	☐ No
A team worker?	☐ Yes	☐ No
Conscientious?	☐ Yes	☐ No

Keen to learn new skills?	☐ Yes ☐ No
Able to understand diagrams?	☐ Yes ☐ No
Able to think on your feet?	☐ Yes ☐ No
Accurate and methodical?	☐ Yes ☐ No
A person with good manual dexterity?	☐ Yes ☐ No
A person with good eyesight?	☐ Yes ☐ No
Anxious to achieve?	☐ Yes ☐ No
Confident about the future of manufacturing?	☐ Yes ☐ No

If you ticked yes to 12 or more then you are obviously well suited to a career in the fast moving world of manufacturing. If you ticked yes to fewer than six, you may need to consider other career options.

Whatever you decide, let me take this opportunity to wish you a very happy and successful career.

CHAPTER 12
FURTHER
INFORMATION

CAREERS ADVICE

Careers Advice
This is a government-funded body designed for everybody which gives advice on jobs and training opportunities.
Website: www.direct.gov.uk/careers advice

Connexions
Tel: 0808 001 3219
Website: www.connexions-direct.com

This a government-funded advisory service with a network of local offices providing advice on jobs and training for young people aged 14–19. If you cannot find a description of a job that suits you in this book, you will find it on the jobs A–Z section of this website.

For information on apprenticeships you should look on the websites www.apprenticeships.org.uk or www.apprentices.gov.uk

Careers Advice in Northern Ireland

Careers and Occupational Information Unit
Department of Employment and Learning, 61 Fountain Street,
Belfast BT1 5EX
Tel: 028 9044 1921
Websites: www.delni.gov.uk and
www.delni.gov.uk/apprenticeships

Careers Advice in Scotland

Careers Scotland
Scottish Enterprise, 150 Broomielaw, Glasgow G2 8LU
Tel: 0845 650 2502
Website: www.careers-scotland.org.uk

Highlands and Islands Enterprise, Cowan House,
Inverness Business and Retail Park, Inverness IV2 7GF
Tel: 01463 234171
Website: www.careers-scotland.org.uk

Careers Advice in Wales

Careers Wales
53 Charles Street, Cardiff CF10 2GD
Tel: 029 2090 6700
Website: www.careerswales.com

SECTOR SKILLS COUNCILS

These are organisations which set the standards and oversee
the training for different manufacturing sectors. Their websites
contain useful information about different industries and the jobs
they offer.

Cogent SSC Ltd
Unit 5, Mandarin Court, Centre Park, Warrington WA1 1EE
Tel: 01925 515200
Website: www.cogent-ssc.com
Sector Skills Council for Chemicals and Pharmaceuticals.

Creative and Cultural Skills
Lafone House, The Leathermarket, Weston Street,
London SE1 3HN
Tel: 020 7015 1800
Website: www.ccskills.org.uk
Sector Skills Council for Design.

Improve Ltd
Providence House, 2 Innovation Close, York YO10 5ZF
Tel: 0845 644 0488
Website: www.improve-skills.co.uk
Sector Skills Council for Food Processing and Drink Manufacture.

Proskills
Century Court, 85B Milton Park, Abingdon OX14 4RY
Tel: 01235 833441
Website: www.proskills.co.uk
Sector Skills Council for building products, glass, coatings, paper,
printing and furniture.

SEMTA
14 Upton Road, Watford WD18 0JT
Tel: 01923 238441
Website: www.semta.org.uk
Sector Skills Council for Science, Engineering and Manufacturing
(including aerospace, automotive and marine manufacturing).

Skillfast
Richmond House, Lawnwood Business Park, Leeds LS16 9RD
Tel: 0113 239 9600
Websites: www.skillfast-uk.org/justthejob;
www.futuretextiles.co.uk; www.canucutit.co.uk;
www.textile-training.com
Sector Skills Council for Clothing, Textiles and Footwear.

EXAMINING BOARDS

Awarding Body Consortium
Robins Wood House, Robins Wood Road, Aspley,
Nottingham NG8 3NH
Tel: 0115 854 1616
Website: www.abcawards.co.uk

City & Guilds of London Institute
1 Giltspur Street, London EC1A 9DD
Tel: 020 7294 2468
Website: www.cityandguilds.com
City & Guilds is the leading provider of vocational qualifications
in the UK. These include NVQs and SVQs.

Edexcel
One90 High Holborn, London WC1V 7BH
Website: www.edexcel.com
An examining body which offers BTEC diplomas and certificates,
including the First Certificate, the National Diploma/Certificate
and the Higher National Diploma/Certificate.

OCR
2 Hill Street, Cambridge CB1 2EU
Tel: 01223 553998
Website: www.ocr.org.uk

TRADE ASSOCIATIONS

Association of the British Pharmaceutical Industry
12 Whitehall, London SW1A 2DY
Tel: 0870 890 4333
Website: www.abpi.org.uk

Become instantly more attractive

To employers and further education providers

Whether you want to be an architect (Construction and the Built Environment Diploma); a graphic designer (Creative and Media Diploma); an automotive engineer (Engineering Diploma); or a games programmer (IT Diploma), we've got a Diploma to suit you. By taking our Diplomas you'll develop essential skills and gain insight into a number of industries. Visit our website to see the 17 different Diplomas that will be available to you. www.diplomainfo.org.uk.

British Apparel and Textile Confederation
5 Portland Place, London W1B 1PW
Tel: 020 7636 7788
Website: www.apparel-textiles.co.uk

British Coating Federation
James House, Bridge Street, Leatherhead KT22 7EP
Tel: 01372 360660

British Glass
9 Churchill Way, Chapeltown, Sheffield S35 2PY
Tel: 0114 290 1850
Website: www.britglass.org.uk

British Printing Industries Federation
11 Bedford Row, London WC1R 4DX
Tel: 020 7915 8300
Website: www.britishprint.com

Chemical Industries Association
Kings Buildings, Smith Square, London SW1P 3JJ
Tel: 020 7834 3399
Website: www.cia.org.uk

Confederation of Paper Industries
1 Rivenhall Road, Swindon SN5 7BD
Website: www.paper.org.uk
Offers courses and qualifications in paper technology.

Intellect
Russell Square House, 10–12 Russell Street,
London WC1B 5EE
Tel: 020 7331 2040
Website: www.intellectuk.org
IT, Telecommunications and Electronics Association.

Society of British Aerospace Companies
Salamanca Square, 9 Albert Embankment, London SE1 7SP
Tel: 020 7091 4545
Website: www.sbac.co.uk

Society of British Motor Manufacturers and Traders
Forbes House, Halkin Street, London SW1X 7DS
Tel: 020 7235 7000
Website: www.smmt.co.uk

PROFESSIONAL ASSOCIATIONS

Chartered Institute of Personnel and Development
151 The Broadway, London SW19 1JQ
Tel: 020 8612 6208
Website: www.cipd.co.uk

Chartered Institute of Purchasing and Supply
Easton House, Easton on the Hill, Stamford PE9 3NZ
Tel: 01780 756777
Website: www.cips.org

Chartered Management Institute
Management House, Cottingham Road, Corby NN17 1TT
Tel: 01536 207307
Website: www.managers.org.uk

Chartered Quality Institute
12 Grosvenor Crescent, London SW1X 7EE
Tel: 020 7245 6722
Website: www.cqi.org

Chartered Society of Designers
5 Bermondsey Exchange, 179–182 Bermondsey Street,
London SE1 3UW
Tel: 020 7357 8088
Website: www.csd.org.uk

Institute of Brewing and Distilling
33 Clarges Street, London W1J 7EE
Tel: 020 7499 8144
Website: www.ibd.org.uk
The IBD offers two awards: General Certificate in Brewing and
Packing; General Certificate in Distilling.

Institute of Leadership and Management
Stowe House, Wetherstowe, Lichfield WS13 6TJ
Tel: 01543 266869
Website: www.i-l-m.com

Institute of Marine Engineering, Science and Technology
80 Coleman Street, London EC2R 5BB
Tel: 020 7382 2600
Website: www.imarest.org

Institute of Operations Management
University of Warwick Science Park, Sir William Lyons Road,
Coventry CV4 7EZ
Tel: 024 7669 2266
Website: www.iomnet.org.uk

Institution of Engineering Designers
Courtleigh, Westbury Leigh, Westbury BA13 5TA
Tel: 01373 622801
Website: www.ied.org.uk

Institution of Engineering and Technology
Michael Faraday House, Six Hills Way, Stevenage SG1 2AY
Tel: 01438 313311
Website: www.theiet.org.uk

REFERENCES

Careers in Fashion and Textiles, Helen Gworek (Blackwell)
Careers in Food and Drink (William Reed Business Media);
    website: www.careersinfoodanddrink.co.uk
Careers in Maritime Engineering (Institute of Marine Engineering)
Careers in the Pharmaceutical Industry (ABPI)
Real Life Guides: Carpentry & Cabinet-Making (Trotman)
Real Life Guides: The Motor Industry (Trotman)
Working in Manufacturing (Connexions)